Material Detectives: M

Let's Look at a Knife and Fork

Angela Royston

www.raintreepublishers.co.uk
Visit our website to find out more information about **Raintree** books.

To order:
Phone 44 (0) 1865 888112
Send a fax to 44 (0) 1865 314091
Visit the Raintree Bookshop at **www.raintreepublishers.co.uk** to browse our catalogue and order online.

First published in Great Britain by Raintree, Halley Court, Jordan Hill, Oxford OX2 8EJ, part of Harcourt Education.
Raintree is a registered trademark of Harcourt Education Ltd.

Editorial: Andrew Farrow and Sarah Chappelow
Design: Jo Malivoire and AMR
Picture Research: Erica Newbery
Production: Duncan Gilbert

Originated by Modern Age
Printed and bound in China by South China Printing Company

10 digit ISBN 1 844 43428 1 (hardback)
13 digit ISBN 978 1 844 43428 2 (hardback)
10 09 08 07 06
10 9 8 7 6 5 4 3 2 1

10 digit ISBN 1 844 43433 8 (paperback)
13 digit ISBN 978 1 844 43433 6 (paperback)
11 10 09 08 07
10 9 8 7 6 5 4 3 2 1

British Library Cataloguing in Publication Data
Royston, Angela
Metal: let's look at a knife and fork – (Material Detectives)
620.1'6
A full catalogue record for this book is available from the British Library

Acknowledgements
The publishers would like to thank the following for permission to reproduce photographs: Harcourt Education Ltd/Gareth Boden pp. **4**, **5**, **17**; Tudor Photography/Harcourt Education Ltd pp. back cover (plate and triangle), **6**, **7**, **8**, **9**, **10**, **11**, **12**, **13**, **14**, **15**, **16**, **18**, **19**, **20**, **21**, **22**, **23** (all), **24**.

Cover photograph of a knife and fork reproduced with permission of Taxi/Getty Images.

Every effort has been made to contact copyright holders of any material reproduced in this book. Any omissions will be rectified in subsequent printings if notice is given to the publishers.

The paper used to print this book comes from sustainable resources.

Some words are shown in bold, **like this**. You can find them in the glossary on page 23.

Contents

What are a knife and fork? 4

How strong are a knife and fork? 6

What sound does metal make? 10

Are a knife and fork smooth or rough? 12

What shape is a knife? 16

What shape is a fork?. 18

How long will a knife and fork last? . . 20

Quiz 22

Glossary 23

Index 24

What are a knife and fork?

You use a knife and fork to eat.

You cut the food with the knife.

Then you put some food in your mouth with the fork.

How strong are a knife and fork?

A knife and fork are very strong.

They are made of a **metal** called steel.

Which of these things are made of metal?

Keys are made of **metal**.

It is not easy to bend metal.

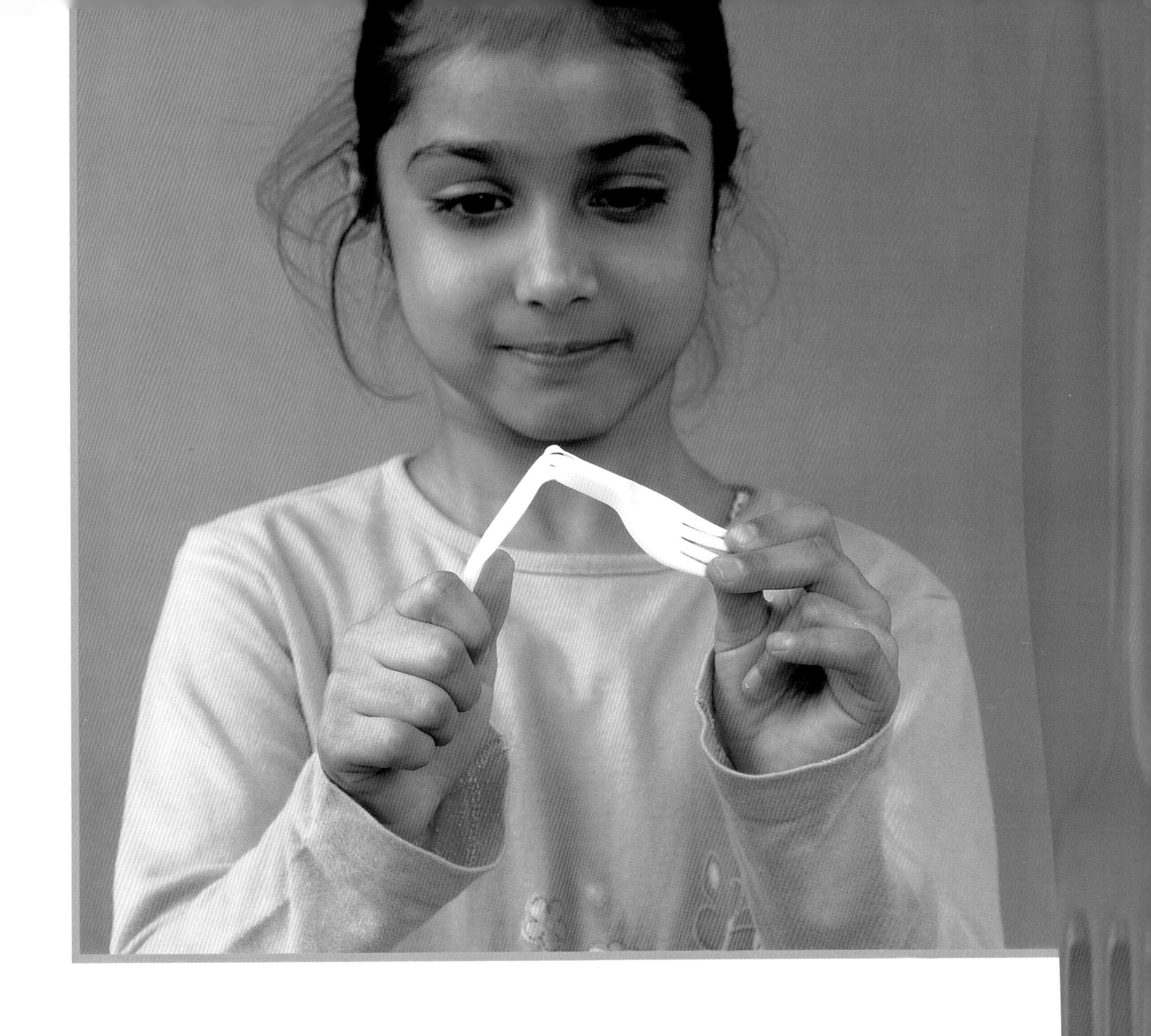

A metal fork is stronger than a plastic fork.

You could not break a metal fork.

What sound does metal make?

Tap a **metal** knife and fork together.

They **ring** because they are both **hard** and metal.

What happens when you tap a metal triangle?

It rings too!

Are a knife and fork smooth or rough?

A knife and fork feel smooth when you touch them.

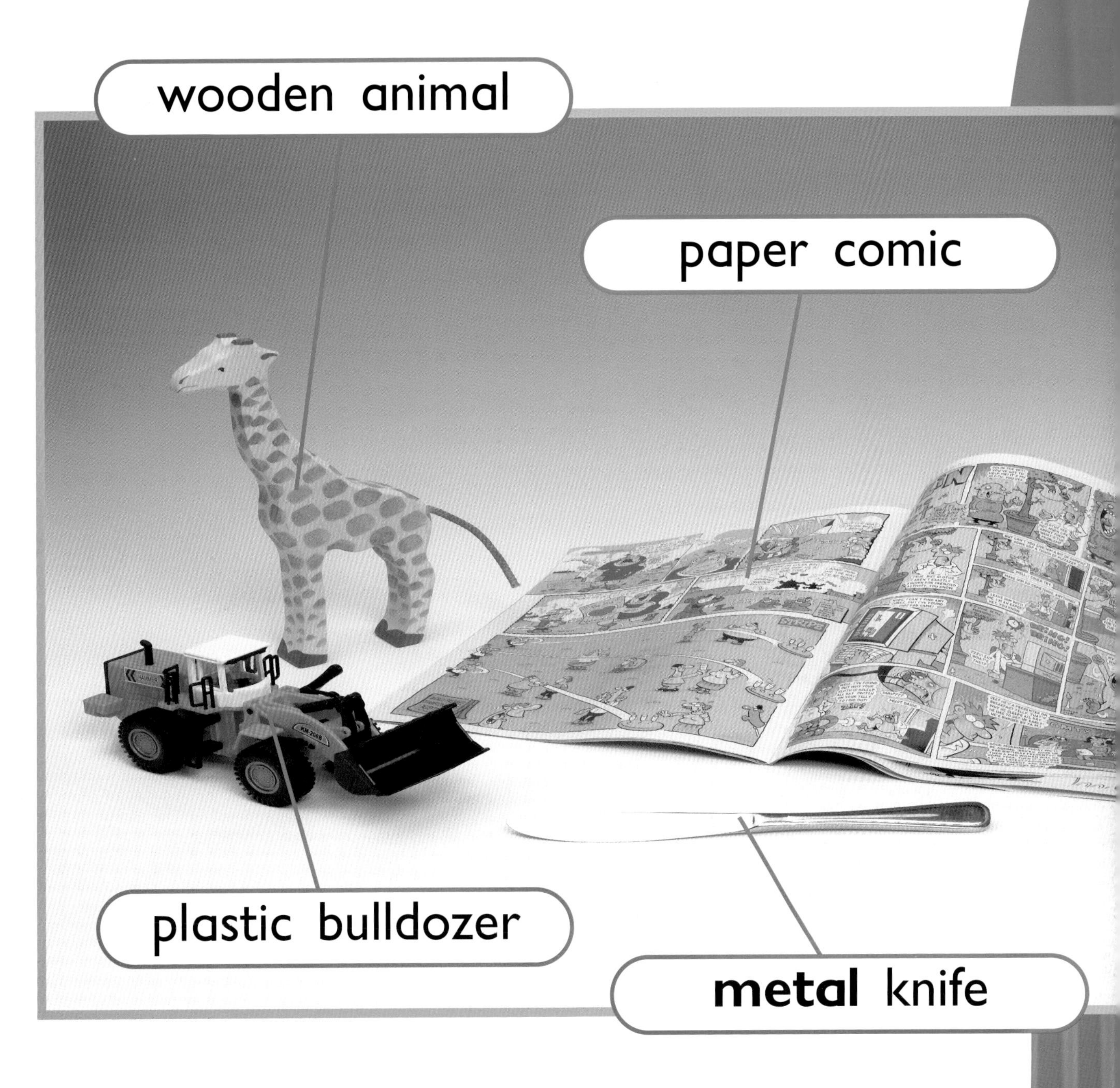

A fork is so smooth it is shiny.

Which of these objects are shiny?

A knife is shiny too.

It is shiny like a mirror.

Smooth things are easy to clean.

It is easy to clean a knife.

What shape is a knife?

A knife is straight.

A knife has a flat **blade** for cutting.

Some knives are very sharp.

Be careful not to cut yourself.

What shape is a fork?

A fork is curved. It has four sharp spikes called prongs.

The prongs hold the food while you cut it with the knife.

How long will a knife and fork last?

Some **metal rusts** when it gets wet.

It slowly turns rough and orange.

Metal knives and forks are made of special steel.

They never rust!

Quiz

This girl can see her face in one of these things.

Is it the wooden spoon or the **metal** fork?

Look for the answer on page 24.

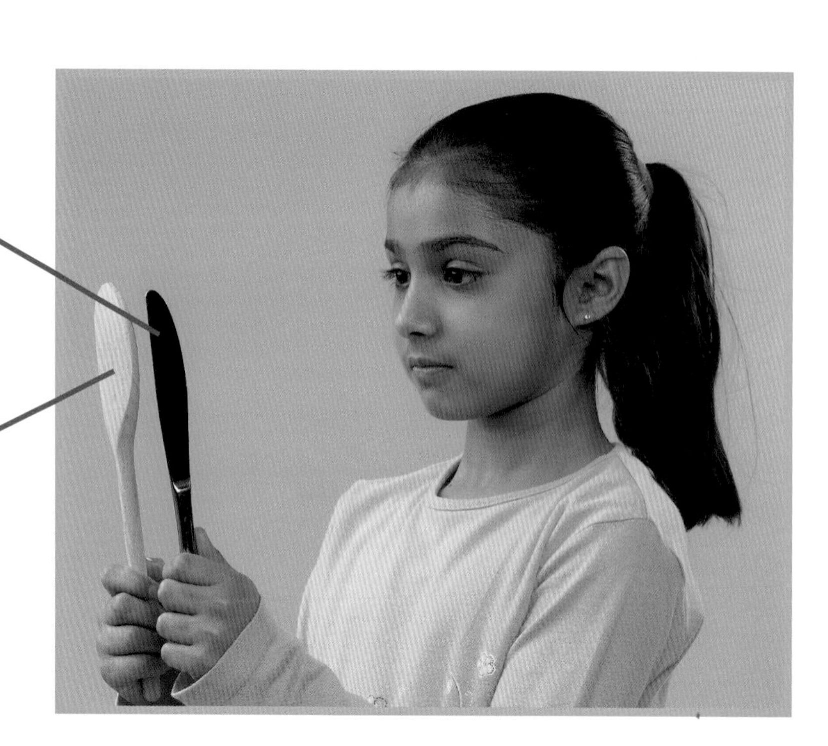

Glossary

blade
flat, sharp part of a knife used for cutting

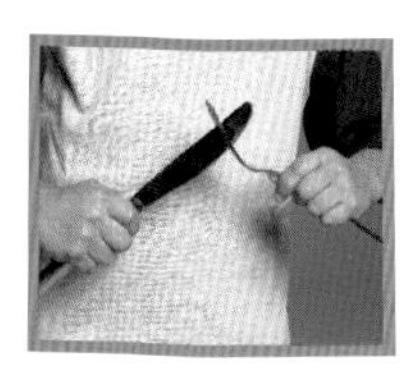

hard
not soft, so you cannot squash it

metal
hard, shiny material

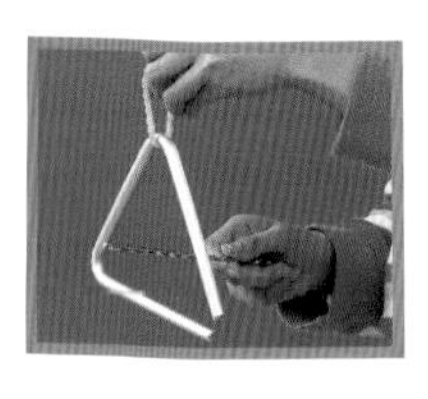

ring
make a sound like a bell

rust
when metal slowly turns orange and crumbles

Index

blade 16
cleaning 15
cut 4, 17, 19
food 4–5
keys 7, 8
metal 6, 7, 8
mirror 14
plastic 9
prongs 18–19
rust 20–21
sharp 17
shiny 13, 14
sound 10
steel 6, 9, 21
touch 12
triangle 11

Answer to quiz on page 22

The girl can see herself in the knife because it is so smooth and shiny.

Note to parents and teachers

Reading for information is an important part of a child's literacy development. Learning begins with a question about something. Help children think of themselves as investigators and researchers by encouraging their questions about the world around them. Each chapter in this book begins with a question. Read the question together. Look at the pictures. Talk about what you think the answer might be. Then read the text to find out if your predictions were correct. Think of other questions you could ask about the topic, and discuss where you might find the answers. Assist children in using the picture glossary and the index to practice new vocabulary and research skills.